BestMasters

Springer awards „BestMasters" to the best master's theses which have been completed at renowned universities in Germany, Austria, and Switzerland.

The studies received highest marks and were recommended for publication by supervisors. They address current issues from various fields of research in natural sciences, psychology, technology, and economics.

The series addresses practitioners as well as scientists and, in particular, offers guidance for early stage researchers.

Melanie Weichselbaumer

Pyridine-functionalized Polymeric Catalysts for CO_2-Reduction

Foreword by Dr. Elisa Tordin, DI Stefanie Schlager
and o.Univ. Prof. Mag. Dr. DDr. h.c. N. S. Sariciftci

 Springer Spektrum

Melanie Weichselbaumer
Linz, Austria

BestMasters
ISBN 978-3-658-10357-6   ISBN 978-3-658-10358-3 (eBook)
DOI 10.1007/978-3-658-10358-3

Library of Congress Control Number: 2015941509

Springer Spektrum

Printed on acid-free paper

Springer Spektrum is a brand of Springer Fachmedien Wiesbaden
Springer Fachmedien Wiesbaden is part of Springer Science+Business Media
(www.springer.com)

Foreword

The increase of the greenhouse gases content in the atmosphere together with the climate change are currently some of the most discussed topics in society and science, raising awareness towards renewable energy. In the past years, a lot of effort and investments have been put into the development of efficient and economically competitive power generation systems such as solar cells or wind turbines. Nevertheless, fossil fuels still provide major part of the energy all around the world. Moreover, not only the generation of power is challenging in the field of renewable energy, but also the devise of reliable and scalable energy storage mechanism is an issue that has to be taken into account. The direct conversion of renewable energy to chemical energy is an appealing alternative since conventional storage and transport of chemicals as solids, liquids and gases are already well developed. A particularly interesting approach is the Carbon Capture and Utilization (CCU) in which (renewable) energy is used to chemically reduce the greenhouse gas carbon dioxide into valuable organic molecules. Such an approach would solve many problems at once. For instance, not only the atmospheric concentration of CO_2 would be kept under control, it would be also recycled and the energy would be chemically stored as high-energy-content molecules. Moreover, renewable energy is not a constant source of energy, thus the chemical conversion into fuel is one of the possible strategies available to utilize sun or wind power continuously and store the provided energy accordingly. Given these premises, the aim of the research described in this thesis concerns the electrochemical carbon dioxide conversion to useful molecules that could be used to replace fossil fuels or as starting materials for chemical synthesis and, eventually, fuel production. Since CO_2 is highly stable, its necessary to lower the activation energy of the reduction reaction and for this purpose, catalysts have to be employed. Possible catalysts range from organic molecules, to metal-organic complexes to biological systems, as bacteria or enzymes. Here we focused on the first two kinds of systems and in particular on the pyridinum and rhenium-bipyridine complex catalysts. Both systems have been investigated as homogeneous catalysts for the reduction of CO_2 to methanol and CO, respectively and they perform well at room temperature and atmospheric pressure. In addition, pyridinium is a simple organic molecule, available in large amount, easily processable and cheap. However, homogeneous catalysts are dissolved in the reaction mixture with the starting materials and, if we think about the possible application of such systems on the big scale, they are not the best option. In fact, a big amount of material is required in order to obtain reasonable yield of final product in a reasonable time range. Moreover, deactivation phenomena may occur in solution, they can hardly be recovered for further use and the difference in solubility between the catalysts and the starting material limits the concentration of one or the other species, leading to dramatic effect on the kinetic of the overall process. The approach described in this thesis depicts a favorable possibility to convert such known homogeneous catalysts into heterogeneous systems by functionalization of a polymeric film grown on an electrode. The catalytic activity of the active sites is preserved and further reusability is achieved after the immobilization of the catalyst. This new approach enables the possible utilization of such catalysts for the big scale CO_2 reduction in industry.

E. Tordin, S. Schlager, N.S. Sariciftci

Acknowledgement

First of all I want to thank my family for their personal and financial support during my studies.

I want to specially thank o.Univ. Prof. Mag. Dr. DDr. h.c. N. S. Sariciftci for giving me the opportunity to work and write on this topic, as well as Dr. Elisa Tordin for supervising and helping me with the practical work. Additionally I want to thank Dogukan H. Apaydin, Stefanie Schlager, Engelbert Portenkirchner, Matthew White, Gottfried Aufischer and Helmut Neugebauer for fruitful discussion and for their help.

Furthermore I want to appreciate for encouragement and friendship the whole LIOS-Team.

THANK YOU!!

Melanie Weichselbaumer

Acknowledgements

[faded, largely illegible text]

Abstract

This study uses pyridine and Rhenium bipyridine functionalized poly-thiophenes as heterogeneous catalysts for carbon dioxide (CO_2) reduction. Carbon dioxide is a greenhouse gas and its atmospheric concentration has increased extremely in the past decades. This alarming trend has to be stopped or, at least, reduced, since greenhouse gases lead to global warming. One way is to use carbon dioxide as a chemical feedstock for higher energy molecules like methane or methanol [5]. Bocarsley reported already in 1994 that pyridinium ion can be used as a homogenous catalyst for CO_2-reduction [34]. In this study, pyridine is linked to a thiophene unit bearing an aliphatic chain to generate 4-(7-(3-thienyl)heptyl)pyridine. Our second approach is to immobilize a [Re(4-Methyl-4'-(7-(3-thienyl)heptyl)-2,2'-bipyridinyl)(CO)₃Cl] on an electrode and reduce carbon dioxide to carbon monoxide. Subsequent electropolymerization on Pt-electrode gives a pyridine functionalised polythiophene and a Rhenium bipyridine functionalised polythiophene. Both polmers are employed as heterogenous catalyst for CO_2-reduction. The electrolysis experiments were carried on for 40 hour and the products analysed by gas chromatography (of headspace and solution) and ionic chromatography.

Zusammenfassung

In dieser Arbeit werden Pyridin- und Rheniumbipyridin-funktionalisierte Polythiophene als heterogene Katalysatoren für die Kohlendioxid (CO_2) Reduktion verwendet. Kohlendioxid ist ein Treibhausgas, dessen Konzentration in der Luft in den letzten Jahren drastisch angestiegen ist. Eine Möglichkeit zur Minderung der CO_2-Konzentration ist die Verwendung von CO_2 als Rohstoff für die Produktion energiereicherer Moleküle wie Methan oder Methanol [5]. Bocarsly hat bereits 1994 berichtet, dass Pyridinium als homogener Katalysator für die CO_2 Reduktion verwendet werden kann [34]. In dieser Studie wurde 3-Bromothiophen mit 1,6-Dibromohexyl umgesetzt und anschließend mit 4-Picolin funktionalisiert, wodurch 4-(7-(3-thienyl)-heptyl)pyridine entsteht. Ein weiteres Material, das in dieser Arbeit untersucht wird, ist [Re(4-Methyl-4'-(7-(3-thienyl)-heptyl)-2,2'-bipyridinyl)(CO)$_3$Cl]. Die Produkte werden elektrochemisch auf einer Pt-Elektrode polymerisiert und als heterogene Katalysatoren für die CO_2 Reduktion eingesetzt. Nach einer 40-stündigen Elektrolyse wurden mittels Gaschromatography (von Headspace und Elektrolytlsung) und Ionenchromatography die Produkte bestimmt.

Contents

1 Introduction   **1**

1.1   Background 1

1.2   Carbon Dioxide: a Greenhouse Gas 1

1.3   Carbon Capture and Storage (CCS) 5

    1.3.1   Carbon Dioxide Separation 6

    1.3.2   Carbon Dioxide Transport 7

    1.3.3   Carbon Storage 8

1.4   Carbon Capture and Utilization (CCU) or Carbon Dioxide
Recycling 9

    1.4.1   Biological Conversion 10

    1.4.2   Photochemical Conversion 12

    1.4.3   Electrochemical Reduction 12

2 Experimental   **17**

2.1   Apparatus 17

2.2   Materials 17

2.3   Setup for Electropolymerization 18

2.4   Setup for Electrolysis 18

2.5   Calculation of the halfway-potential 19

3 Results and Discussion   **23**

3.1   Chemical Synthesis 23

    3.1.1   Synthesis of 3-(6-bromohexyl)- thiophene 23

    3.1.2   Synthesis of 4-(3-thienylheptyl)pyridine 23

    3.1.3   Synthesis of 4-Methyl-4′-(3-thienylheptyl)-
2,2′-bipyridine 25

    3.1.4   Synthesis of [Re(4-Methyl-4′-(7-(3-thienyl)heptyl)-2,2′-
bipyridinyl)(CO)$_3$Cl] 26

3.2   Electropolymerization 28

    3.2.1   Electropolymerization of Pyridine functionalized poly-
(thiophene) (PP1) 28

3.2.2   Electropolymerization of 4-Methyl-4′-(7-(3-thienyl)-
heptyl)-2,2′- bipyridine 32

3.2.3   Electropolymerization of Rhenium-bipyridine complex functionalized thiophene (RBPP2) 35

3.3   Characterization of the polymeric films 35

3.3.1   Characterization of PP1 film in acetonitrile 35

3.3.2   Characterization of PP1 film in 0.5 M KCl 37

3.3.3   Characterization of RBPP1 film in propylen carbonate 40

3.3.4   Characterization of RBPP2 film 41

3.4   Electrolysis 42

3.4.1   Electrolysis of PP1 film in acidified 0.1 M TBAPF$_6$
acetonitrile solution 42

3.4.2   Electrolysis of PP1 film in 0.5 M water KCl solution . 44

3.4.3   Electrolysis of RBPP2 film in 0.1 M TBAPF$_6$ in acetonitrile 44

4 Conclusion   **47**

5 Appendix   **49**

6 Bibliography   **55**

List of Figures

1   Temperature chances over the last century [8]. 3

2   Correlation of Earth's surface temperature and the amount of carbon dioxide [8]. 4

3   CO_2-measurement on Mauna Loa [9]. 4

4   CO_2-separation principles [15]. 7

5   Geological storage options [44]. 8

6   Sunlight-driven photocatalytic conversion of carbon dioxide to hydrocarbons [32]. 11

7   Chemical structure of (2,2'-bipyridyl)Re(CO)$_3$Cl (1) and (5,5'-bisphenylethynyl-2,2'-bipyridyl)Re(CO)$_3$Cl (2) [39]. 13

8   Qualitative reaction scheme for carbon dioxide conversion [24]. 13

9   chemical structure of 3-(6-Bromohexyl)thiophene: **compound 1** 17

10   chemical structure of 4-Methyl-4'-(3-thienylheptyl)-2,2'-bipyridine: **compound 2** 18

11   chemical structure of 4-(3-Thienylheptyl)-pyridine: **compound 3** 18

12   chemical structure of [4-Methyl-4'-(3-thienylheptyl)- 2,2'-bipyridinyl)Re(CO)$_3$Cl]: **compound 4** 19

13   One-compartment cell for purging with N$_2$ (A) and during electropolymerization a closed system (B). Cells contain a working electrode (WE), a reference electrode (RE) and a counter electrode (CE) [39]. 19

14   H-cell for electrolysis. The cell contains a WE (Pt-plate with polymeric film) and a Ag/AgCl-quasi RE on the one side of the cell and on the other side a CE (Pt-plate). 20

15   Cyclic voltammogram measured of Ferrocene with a scan-rate of 10 mVs^{-1} in 0.1 M TBAPF$_6$ in acetonitrile. 21

16   reaction scheme for synthesis of 3-(6-bromohexyl)thiophene 24

17   reaction scheme for synthesis of 4-(3-thienylheptyl)pyridine 24

18 reaction scheme for synthesis of 4-methyl-4′-(3-thienylheptyl)-2,2′-bipyridine 26

19 reaction scheme for synthesis of [4-methyl-4′-(3-thienylheptyl)-2,2′-bipyridinyl)Re(CO)$_3$Cl] 27

20 Scan from 0 mV to 1600 mV and back to 0 mV of [Re(4-Methyl-4′-(7-(3-thienyl)heptyl)-2,2′-bipyridinyl)(CO)$_3$Cl] monomer in 0.1 M TBAPF$_6$ in propylencarbonate and after adding 10 % of BFEE with a scanrate of 100 mVs^{-1}. 29

21 Potentiodynamic electropolymerization of 4-(3-Thienylheptyl)-pyridine. The polymerization is done in an one compartment-cell where a Pt-plate was used as working electrode. The counter electrode was also a Pt-plate and an Ag/AgCl-wire was used as quasi-reference electrode. 30

22 UV-vis measurement of pyridine functionalized poly- (thio-phene) film on glass-ITO. 31

23 Potentiodynamic electropolymerization of 4-Methyl-4′-(7-(3-thienyl)heptyl)-2,2′-bipyridine with a scan rate of 50 mVs^{-1} in 0.1 M TBAPF$_6$ in propylen carbonate with 10% BFEE. . . . 33

24 Complexation of poly-4-methyl-4′-(3-thienylheptyl)-2,2′-bipyri-dine film with Re(CO)$_5$Cl in hot toluene. 34

25 Potentiodynamic electropolymerization of [Re(4-Methyl-4′-(7-(3-thienyl)heptyl)-2,2′-bipyridinyl)(CO)$_3$Cl]. The poly-merization is done in an one compartment-cell where a Pt-plate was used as working electrode. The counter electrode was also a Pt-plate and an Ag/AgCl-wire was used as quasi-reference electrode. 36

26 Cyclic voltammogram of polymeric-4-(3-Thienylheptyl)-pyridine-film measured with a scanrate of 50 mVs^{-1}. The polymerization is done in an one compartment-cell where a Pt-plate covert with the polymer was used as working electrode. The counter electrode was also a Pt-plate and an Ag/AgCl-wire was used as quasi-reference electrode. 0.1 M TBAPF$_6$ in acetonitrile is used as electrolyte. 38

27   Cyclic voltammogram of polymeric-4-(3-Thienylheptyl)-
     pyridine-film measured with a scanrate of 1 mVs^{-1}. The
     polymerization is done in an one compartment-cell where
     a Pt-plate covert with the polymer was used as working
     electrode. The counter electrode was also a Pt-plate and an
     Ag/AgCl-wire was used as quasi-reference electrode. 0.1 M
     TBAPF$_6$ in acetonitrile is used as electrolyte. 39
28   Cyclic voltammogram of polymeric-4-(3-Thienylheptyl)-
     pyridine-film measured with a scanrate of 50 mVs^{-1}. The
     polymerization is done in an one compartment-cell where a
     Pt-plate was used as working electrode. The counter elec-
     trode was also a Pt-plate and an Ag/AgCl-wire was used as
     quasi-reference electrode. 0.5 M KCl is used as electrolyte. . 40
29   Cyclic voltammogram measured of the Re-complexed poly-
     meric film with a scanrate of 10 mVs^{-1} in 0.1 M TBAPF$_6$ in
     propylen carbonate 41
30   A proposed mechanism for CO$_2$-reduction with pyridinium
     to methanol [36]. 43
31   Characterization of [Re(4-Methyl-4'-(7-(3-thienyl)heptyl)-2,2'-
     bipyridinyl)(CO)$_3$Cl] film in 0.1 M TBAPF$_6$ in acetonitrile. . . 45
32   40 hour electrolysis of pyridine functionalized poly(thiophene)
     in 0.1 M TBAPF$_6$ in acidified acetonitrile. Holdvalue at a po-
     tential of -1200 mV. 49
33   40 hour electrolysis of Rhenium-bipyridine complex func-
     tionalized poly(thiophene) in 0.1 M TBAPF$_6$ in acetonitrile.
     Holdvalue at a potential of -1800 mV. 50
34   ^{1}H-NMR of 3-(6-Bromohexyl)-thiophene in CDCl$_3$ 51
35   ^{1}H-NMR of 4-3-thienylheptyl-pyridine in CDCl$_3$ 52
36   ^{1}H-NMR of 4'-3-thienylheptyl-2,2'-bipyridine in CDCl$_3$ . . . 53
37   ^{1}H-NMR of 4-Methyl-4'-(3-thienylheptyl)-2,2'-bipyridinyl)-
     Re(CO)$_3$Cl 54
All figures can be accessed on www.springer.com under the author's name
and the book title.

1 Introduction

1.1 Background

One of the greatest problems we have been facing is the rise of the Earth's surface temperature, also known as global warming. The main reason for this is the increase of the concentration of greenhouse gases, such as carbon dioxide (CO_2), in the atmosphere which is mostly related to human technological activities. The climate change will not only influence social life and economy but also environmental systems. Therefore one might say that global warming affects everything and everybody on the planet [1]. Carbon dioxide occurs naturally in the Earth's carbon cycle, which is the natural circulation of carbon among the atmosphere, hydrosphere and lithosphere. Anthropogenic CO_2 strongly affects the carbon cycle by adding more CO_2 to the atmosphere and by influencing the capability of natural sinks, like forests, to remove it. CO_2 comes from a variety of natural sources but human-related emissions (like combustion of fossil fuels) are known to be responsible for the increase that has occurred since the industrial revolution [2].

1.2 Carbon Dioxide: a Greenhouse Gas

Greenhouse gasses absorb thermal radiation and re-emits it in all directions. Part of it is emitted back to the Earth's surface and lower atmosphere, inducing the increase of the temperatures and the so called "global warming" [3]. Carbon dioxide is constantly being exchanged among the atmosphere, ocean and land surface because it is not only produced but also absorbed by many microorganisms, plants and animals and the emission and removal of natural carbon dioxide by natural processes tend to balance [2]. Industrial revolution, burning forest lands, mining, burning coal and the development of the society as we know it, led to an increase of the emission of greenhouse gases and to the increase of the Earth's temperature. Effects of global warming could be: extreme weather, plant and animal extinctions, rising sea levels and major shifts in climate [4]. Among

1

all the possible sources of anthropogenic CO_2, the one that contributes most heavily on the overall amount of this gas is the use of carbon-based fossil fuels in human activities [5, 6]. Currently, carbon-based fossil fuels represent a majority of the world's energy sources and it will be the same in near future [5].

The discussion on climate change started already in the 19th century. A seminar paper about possible influences on Earth's heat balance was published by John Tyndall, a British physicist, describing the phenomenon that is nowadays known as the greenhouse effect.

Many scientist tried to calculate the increase of the average global surface temperature [1]. Arrhenius, a Swedish chemist, calculated that doubling the amount of carbon dioxide in the atmosphere could produce a 3.3°C increase in the Earth's surface temperature. Later Callender estimated a 1.1°C, while Plass estimated a 3.6°C increase. Those calculations correlate quite well with the calculation made with todays mathematical climate models that estimate an increase of 1.9-5.2°C if the concentration of CO_2 doubles. Experimentally it has been observed that, in the last century the carbon dioxide concentration increased around 30% and the temperature increased by 0.3-0.6°C. Plass correctly estimated the increase of the concentration of carbon dioxide but he overestimated the increase of temperature. As the task is very complex, the temperature estimation is within an acceptable margin [1].

Scientists observed global warming over the past few years. Figure 1 shows the temperature changes over the last century. With indirect measurements of climates such as analyzing ice cores, tree rings and ocean sediments scientist were able to find out how the temperature on earth changed over thousands of years. When snow sinters and is transformed into ice, some air is trapped in to the pores of the newly formed ice. By crushing a sample of it under vacuum without melting, carbon dioxide concentration can be measured by gas chromatography. Until the past century, natural factors caused atmospheric carbon dioxide concentration to vary between 180 and 300 parts per million (ppm). Warmer periods match to periods of relatively high carbon dioxide concentrations. In

2

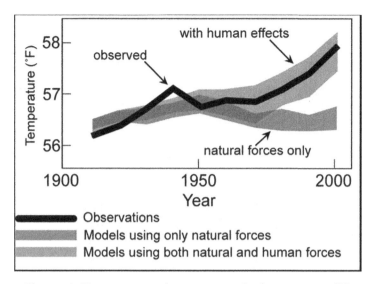

Figure 1: Temperature chances over the last century [8].

Figure 2 one can see the direct correlation between the temperature and the amount of carbon dioxide [7, 8].

David Keeling started to record direct measurements of carbon dioxide in the atmosphere in 1958. The data are measured on Mauna Loa (Hawaii) as a mole fraction in dry air which is defined as the number of molecules of carbon dioxide divided by the number of all molecules in air, including carbon dioxide, after water vapour has been removed. The mole fraction is expressed as parts per million (ppm). The measurements of atmospheric carbon dioxide on Mauna Loa form the longest record of direct measurements of CO_2 in atmosphere [9].

Figure 3 shows the amount of carbon dioxide in atmosphere measured at Mauna Loa between 1958 and the beginning of 2014. The amount of CO_2 increased every year and is currently at a record high. In the past 50 years the amount of atmospheric CO_2 rose about 80 ppm. The sarrated curve shows the not-corrected carbon dioxide concentration data and the black curve shows the seasonally corrected data [9]. In 1976 two major features of the measurements in Mauna Loa were reported, namely the seasonal oscillation and a long term increase [11].This oscillation originates from the

3

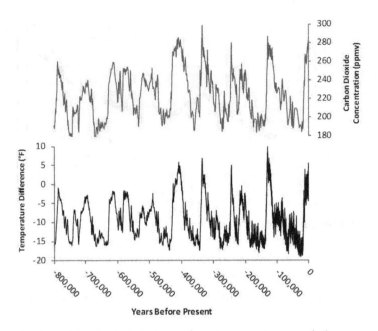

Figure 2: Correlation of Earth's surface temperature and the amount of carbon dioxide [8].

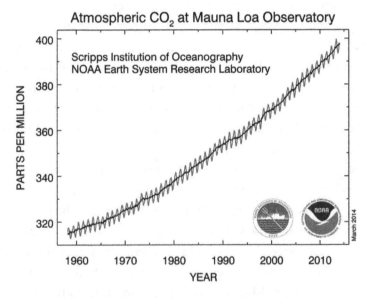

Figure 3: CO_2-measurement on Mauna Loa [9].

4

greater extent of landmass in the northern hemisphere and its vegetation. In spring and summer less CO_2 is measured in the atmosphere because plants convert it to plant material through photosynthesis [6].

In the last century human activities have increased the concentration of the greenhouse gases in the atmosphere. Those gases are considered to have a significant impact on the climate because they cause a global warming [12]. The changes in the average temperature on Earth also cause the melting of glaciers and therefore a rising of water level. This is seen as a major problem especially in coastal areas. The atmospheric concentration of carbon dioxide in the year 2030 is predicted to double in comparison to the level in 1990. At the moment there are several possibilities for reducing the emission but another problem is that CO_2 concentration, temperature and sea level will continue to rise long after emissions are reduced. Therefore it is very hard to make long term predictions of the effects that greenhouses gases will have on the climate of Earth. Nevertheless it is necessary to make measurements and predictions to avoid reaching the "point of no return" [5].

There are several different approaches to reduce carbon dioxide emission. The most important possibilities are carbon capture and storage (CCS) and the use of biomass [12].

One additional method is carbon dioxide utilization, in which carbon dioxide is recycled and can therefore be re-used as feedstock. It is important to point out some of those methods still need to be developed in order to become more effective and be used for applications on the large scale [5].

1.3 Carbon Capture and Storage (CCS)

Carbon dioxide from fossil fuel combustion is a major contributor to climate change and one step to reduce emission is to capture carbon dioxide right after combustion [13]. Carbon Capture and Storage begins with the capture of carbon dioxide at an industrial plant or fossil fuel power plant [14]. The process consists mainly of three steps: separation of carbon dioxide in a concentrated form at the power plant, transportation to a secure location where it can be <u>stored</u> underground [16]. Carbon capture and storage, or carbon capture and sequestration, is widely studied in Europe and in the USA [12].

1.3.1 Carbon Dioxide Separation

CO_2 can be separated from the other gasses mainly in three ways, which are shown in Figure 4:

- Converting the discharge of an industrial process to a pure (or near-pure) CO_2 stream. Carbon dioxide is, then, removed from the waste gas by chemical absorption on an aqueous amine solution at relatively low temperatures, in the so called *post-combustion capture*. In the regeneration process the CO_2 is dried, compressed and afterwards transported to a geological storage site.

- A pure (or near-pure) CO_2 stream, from an existing industrial process is used to form "synthesis gas". Water is added and the mixture has to pass through a series of catalysts for the "water-gas shift" reaction. CO and H_2O reacts to CO_2 and H_2. This process is also known as *pre-combustion capture*.

- Direct air capture into a pure CO_2 stream. In *oxyfuel combustion* plants the main step of the process is the separation of oxygen and nitrogen. The fuel is then burned in a mixture of oxygen and recycled gas of the powerplant. The products are mainly CO_2 and water vapour, which can be separated and cleaned easily [17, 18].

The post-combustion process is an "end-of-pipe" procedure and could therefore be integrated in existing industrial processes. Nevertheless it is connected to high costs and energy losses. At the moment the only system that is commercially available is CO_2 separation that uses chemical absorption but it is still necessary to conduct further research. The pre-combustion has a lower energy requirement but the plants and their operations are very complex and further development is necessary before

6

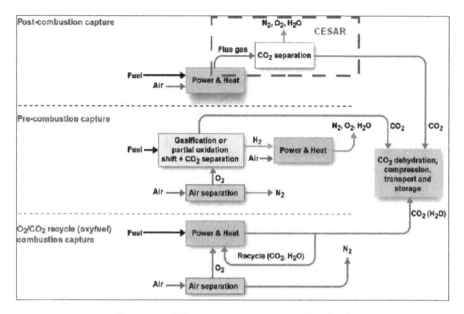

Figure 4: CO_2-separation principles [15].

the technology will be commercially available. In the oxyfuel process, CO_2 is very concentrated and the flue-gas stream is processed more easily than in the other processes. A disadvantage is that the production of pure oxygen and also the liquefying of air requires a lot of energy [16].

1.3.2 Carbon Dioxide Transport

The carbon dioxide has to be compressed after separation in order to be transported. The goal is to find the cheapest and most efficient way to transport carbon dioxide [16]. Rail and truck are not an option for transportation because of their low capacity and high costs.Ships and pipelines are better options. Pipelines are employed worldwide for oil and other hazardous substances, but corrosion-resistant materials are needed for CO2 transport. A relevant environmental aspect of transport via pipelines is the risk of leakages. Carbon dioxide is not toxic but it can be dangerous for living things if the local concentration is too high.

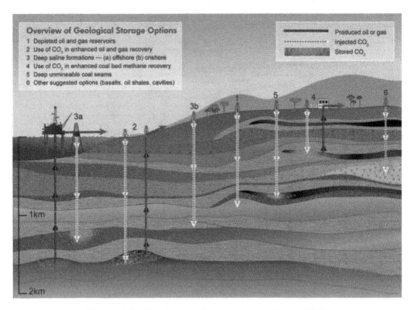

Figure 5: Geological storage options [44].

1.3.3 Carbon Storage

There are several different possibilities for the storage of carbon dioxide. The first possibility is injecting it as a supercritical fluid to deep geological subsurface rock formations where it can be stored safely and permanently as shown in Figure 5 [14, 19, 20]. A second possibility is to use carbon dioxide for enhancing oil and gas extraction. The pressure underground is increased and raw oil and gas products, which would be otherwise left behind, are pushed out to the surface [21, 22]. In such a way, additional oil and gas are recovered and the overall costs of storage decrease [16]. The disadvantage is that this method contributes more to the problem as the extracted oil will be burned to generate more CO_2.

In California carbon dioxide is also used for carbonation of brines [21]. The porous sediments are filled with salty water and the pore space can be used to take up carbon dioxide. In domed structures sideways for migration of carbon dioxide are limited. Saline aquifers offer the biggest volume potential for storing carbon dioxide [16].

8

One possibility is also to inject and dissolve carbon dioxide in seawater. As carbon dioxide is denser than water it is expected to form a sort of "lake" below the water. Ocean storage is still in the research phase [21, 23]. Placing CO_2 in the deep ocean would isolate it from the atmosphere for several centuries. However, over long times the ocean and atmosphere will equilibrate. Therefore injection of carbon dioxide to the ocean reduces atmospheric CO_2 for several centuries but not on the millennial time scale. Moreover, the concentration of CO_2 in water would slowly increase, changing the pH and affecting aquatic life-forms [23]. In artificial mineral carbonation carbon dioxide should react with a source rock (usually silicates) to form carbonates. This natural process of rock weathering is normally extremely slow and it is not known if it can become technically manageable [16].

1.4 Carbon Capture and Utilization (CCU) or Carbon Dioxide Recycling

This thesis will focus on another, more convenient approach to the CO_2 problem. Instead of storing the CO_2 underground, one can use it as chemical feedstock for the production of more energy rich molecules.

In order to minimize the effects of carbon dioxide as a greenhouse gas in the atmosphere one needs to find a way to reduce emissions by reducing the use of fossil fuel and develop efficient renewable energy source [24]. Carbon dioxide from fossil fuel could be used as a raw material in chemical industry to produce commercial products and energy rich molecules [25]. Carbon dioxide is one of the cheapest and most abundant carbon-containing raw material in the world and it can be easily seen as possible C_1-building block for carbon chemistry. The main drawback is that many reactions are energetically unfavorable [1]. In the Greenmethanol Synthesis, waste carbon dioxide from industry is used to produce methanol. The hydrogen which is necessary for the reaction is generated by electrolysis of water [20]. Langanke *et al* describe the use of carbon dioxide as a feedstock for polyurethane production. One can produce alternating polycarbonates or polyethercarbonates by copolymerizing carbon dioxide and epoxides [26]. Various metal centers bound to rigid ligands are used as single-site catalysts [27].

If methanol can be produced out of carbon dioxide in an energetically efficient way and on the large scale it could replace oil and gas as a fuel and as a chemical raw material. Moreover, because of the shortage of fossil fuels this new process of converting carbon dioxide should be of great interest since it opens new possibilities for cheap and "clean" energy recovery [28].

1.4.1 Biological Conversion

Carbon dioxide can be used as food for algae in solar-active membranes [20]. Algae are grown in artificial ponds which are fertilized with carbon dioxide. Under these conditions it is possible to grow microalgaes, harvest the biomass and convert it to food, or fuel. Therefore fuel-generated carbon dioxide is avoided. Moreover, if algae are fed to certain animals, they will be converted to methane, which is which is an even stronger greenhouse gas. On the other hand if the biomass is converted to biofuel it replaces fossil fuel and therefore the emission of fossil fuel-generated carbon dioxide is avoided [25]. When compared to plants, algae are favourite since they do not have roots, stems and leafs and the entire biomass can be used for many purposes. Another advantage is that they can be cultivated in seawater and so they do not compete with resources of conventional agriculture [49]. Therefore, this approach has a huge potential for the recycling and reduction of carbon dioxide emissions [5].

Another possibility is the reduction of carbon dioxide with hydrogen, where carbon dioxide is taken from power plants and hydrogen is provided from water [28]. Carbon dioxide is dissociated thermally to carbon monoxide (CO) and oxygen and the Fischer-Tropsch process is used to convert the carbon monoxide to hydrocarbons. The required temperatures of more than 2000°C can be achieved with a mirror which focuses sunlight into the reactor [20].

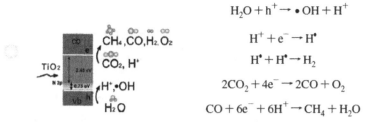

$$H_2O + h^+ \rightarrow \bullet OH + H^+$$

$$H^+ + e^- \rightarrow H^\bullet$$

$$H^\bullet + H^\bullet \rightarrow H_2$$

$$2CO_2 + 4e^- \rightarrow 2CO + O_2$$

$$CO + 6e^- + 6H^+ \rightarrow CH_4 + H_2O$$

Figure 6: Sunlight-driven photocatalytic conversion of carbon dioxide to hydrocarbons [32].

Y. Lu *et al* reported about an enzymatic approach to convert carbon dioxide. High yields and selectivity under milder conditions are the advantages of this new method. The reduction from CO_2 to formic acid is catalyzed by formate dehydrogenase (FateDH) which is encapsulated in an alginate-silica hydride gel. Reduced nicotinamide adenine dinucleotide (NADH) acts as a terminal electron donor for the enzymatic reaction, which is shown in equation (1). In order to increase the enzyme stability and reduce the enzyme cost, the enzymes are immobilized in a Calcium alginate (Ca-alginate) for the reaction [29].

$$CO_2 + 2\,NADH \xrightarrow{FateDH} HCOOH + 2\,NAD^+$$

(1)

Xu *et al* present an adapted system of enzyme-catalyzed reduction. By co-encapsulating three dehydrogenases in a novel alginate-silica (ALG-SiO_2) matrix, the problem of high enzyme-leakage was solved. In addition a high yield of methanol was achieved [30]. Reda *et al* used a tungsten containing formate dehydrogenase enzyme (FADH1) as a heterogeneous catalyst for the same purpose. Several advantages such as reversibility of electrocatalysis, only small overpotential are necessary and only one product - formate - is formed [31]

1.4.2 Photochemical Conversion

One question is still open: how carbon dioxide can be activated into a reactive form? It is possible to activate it with an alternative energy form such as sunlight, and promote an artificial photosynthetic cycle which turns CO_2 into organic substances [5]. Activation of CO_2 requires high energy photons so it is necessary to use an electron shuttle such as a semiconductor. In p-type semiconductors, for instance, light can be used to promote electrons from the valence band to the conduction band and used to reduce CO_2. The ideal case would be to convert them to hydrocarbons. A very promising cycle, which contains only carbon dioxide and water is shown in Figure 6.

Graetzel reported that photolysis of particulate dispersions of TiO_2 loaded with a ruthenium dye (as photosensitiser) in aqueous solution reduces carbon dioxide to methane [33]. If an aqueous dispersions of titania loaded with a copper-containing dye is used for the reaction, methanol is found as a major reduction product. The solubility of CO_2 in water is, however, quite low when compared to organic solvents [33].

Portenkirchner *et al* reported the use of modified rhenium complexes for the photocatalytic reduction of carbon dioxide in acetonitrile. Figure 7 shows the modified rhenium-compounds which was subject of the study. The extended-electron system increases the optical absorption, improving the photocatalytic applications of the materials. It is important to point out that both compounds are used as homogeneous catalyst for reduction of CO_2 to CO. Experiments over several hours only yielded very low CO formation, therefore further investigation is needed to be done [39]. Inspired by this results, in this thesis we describe the use of rhenium-complexed functionalised polythiophenes as heterogeneous catalysts for electrochemical CO_2-reduction.

1.4.3 Electrochemical Reduction

The electrochemical conversion of CO_2 has promised to help overcome several of the challenges facing the implementation of carbon-neutral en-

12

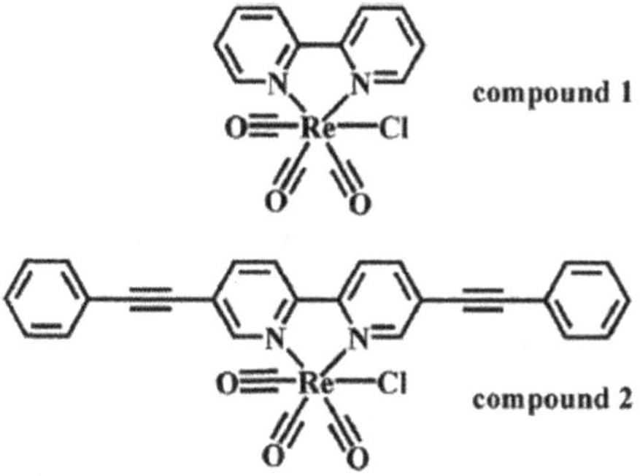

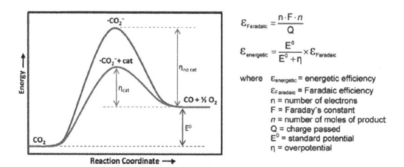

Figure 7: Chemical structure of (2,2'-bipyridyl)Re(CO)$_3$Cl (1) and (5,5'-bisphenylethynyl-2,2'-bipyridyl)Re(CO)$_3$Cl (2) [39].

$$\varepsilon_{Faradaic} = \frac{n \cdot F \cdot n}{Q}$$

$$\varepsilon_{energetic} = \frac{E^0}{E^0 + \eta} \times \varepsilon_{Faradaic}$$

where $\varepsilon_{energetic}$ = energetic efficiency
$\varepsilon_{Faradaic}$ = Faradaic efficiency
n = number of electrons
F = Faraday's constant
n = number of moles of product
Q = charge passed
E^0 = standard potential
η = overpotential

Figure 8: Qualitative reaction scheme for carbon dioxide conversion [24].

13

ergy sources because it provides a means of storing renewable electricity in a convenient, high energy-density form [24]. Research has shown that the electrochemical reduction of carbon dioxide can produce a variety of organic compounds such as formic acid, carbon monoxide, methane and ethylene. Those products can be used as a chemical feedstock for chemical synthesis or they can also be converted into hydrocarbon fuels. This field of research is of great interest because carbon dioxide can be recycled and therefore the accumulation in atmosphere is reduced. In addition renewable hydrocarbons are produced and electrical energy can be stored in chemical form. Real world application of this electrochemical process will happen only when high energy efficiency and rapid reaction rates can be demonstrated. The energetic efficiency is a critical parameter and defines the energy cost of the process. High energetic efficiency can be achieved with a combination of high selectivity (Faradic or current efficiency) and low overpotentials as shown in Figure 8. Catalysts and electrolytes can lower the energy of the intermediate, improving the energetic efficiency of the conversion. To avoid confusion ε is used for the Faradaic and energetic efficiencies an η is used for the overpotential [24].

Electrochemical systems for the reduction of carbon dioxide to methanol or methane were developed around 1990 [34]. Those systems had an excellent product yield but they needed a high electrode overpotential with various electrodes, which inevitably leads to a highly energetically inefficient process. The problem one has to face in aqueous solution is that the reduction of carbon dioxide is in competition with the reduction of protons or water to molecular hydrogen. Different approaches can be followed to avoid the use of high overpotential such as the use of catalysts (or electrocatalysts) that introduce new reaction paths characterised by lower energy of the intermediates. The use of redox active enzymes, Prussian-blue-modified electrodes, molecular redox mediators as well as polypyridyl transition metal complexes are good steps in the right direction but some overpotential is still necessary [34]. In 1994, Bocarsly reported that the addition of pyridinium ions to an aqueous electrolyte provides the environment which is necessary to reduce carbon dioxide to methanol with few hundreds of mV of overpotential [34]. In this thesis a similar idea is described but in order to reduce the costs of the catalyst the idea was to immobilize the pyridinium ions on a Pt-electrode so they can be reused easily.

2 Experimental

2.1 Apparatus

Electropolymerization, electrolysis as well as cyclic voltammetry were performed with the Jaissle Potentiostat-Galvanostat IMP 83PC-10. The headspace analysis was performed with the Thermo Scientific Trace GC Ultra Gas Injection Gas Chromatography while liquid gas chromatography with the Thermo Scientific Trace 1310 Liquid Injection Gas Chromatography. Ions analysis was performed with the Thermo Scientific Dionex Cap-IC. Thickness measurements were performed with DektakXT from Bruker.

2.2 Materials

Pyridinium as well as Re-complexed bipyridines have been reported to work as homogeneous catalyst for CO_2 reduction [34, 40]. As heterogenous catalysts can be reused easily our goal was to immobilize those materials on a Pt-electrode. The idea is to functionalize and electropolymerize thiophene to get an heterogenous catalyst. In this thesis the synthesis of 3-(6-bromohexyl)thiophene (9) and 4-Methyl-4'-(3-thienylheptyl)-2,2'-bipyridine (10) is described. Compound 1 is used as precursor for the synthesis of 4-(3-thienylheptyl)pyridine (11) and compound 2 is used as precursor for the synthesis of [Re(4-Methyl-4'-(7-(3-thienyl)heptyl)-2,2'-bipyridinyl)(CO)$_3$Cl] (12. The pyridine-functionalized and Re-complexed-functionalized thiophenes are electropolymerized, characterized and tested for CO_2 reduction.

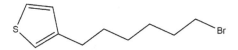

Figure 9: chemical structure of 3-(6-Bromohexyl)thiophene:
compound 1

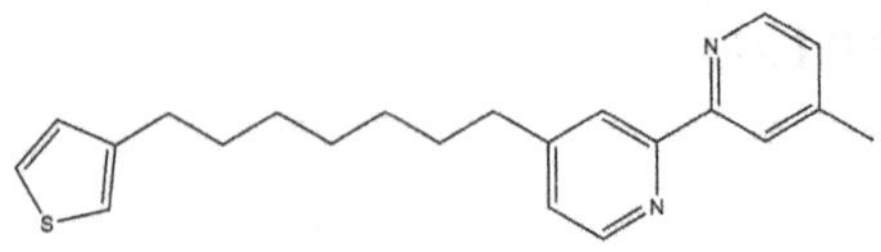

Figure 10: chemical structure of 4-Methyl-4'-(3-thienylheptyl)-2,2'-bipyridine: **compound 2**

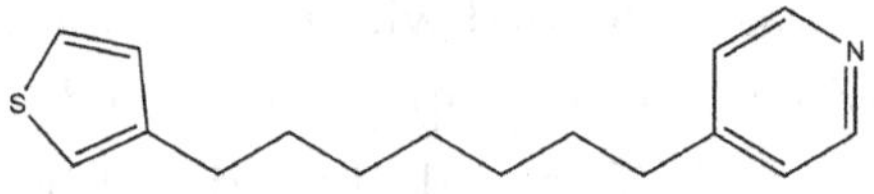

Figure 11: chemical structure of 4-(3-Thienylheptyl)-pyridine: **compound 3**

2.3 Setup for Electropolymerization

The polymerization is done in a three-electode system in a one-compartment cell, which is shown in Figure 13. The working electrode and the counter electrode are Pt-plates. An Ag/AgCl-wire is used as quasi-reference electrode. In order to see if the film is working for CO_2-reduction several cyclic voltammetry measurements are performed in an one-compartment cell using the polymer covered Pt-plate as working electrode, a Pt-plate as counter electrode and an Ag/AgCl-wire as a quasi-reference electrode. If the film does not cover the entire Pt-plate it is important that only the film dips to the solution, so one can be sure that the characteristics of the polymer are tested.

2.4 Setup for Electrolysis

For electrolysis a three-electrode system in a two-compartment H-cell is used, which is shown in Figure 14, is used. On one side of the cell the WE (Pt-plate with polymeric film, where only the film dips in to the solution) and an Ag/AgCl-wire as RE are fixed. On the other side of the cell the CE (Pt-plate), which faces the WE, is fixed. It is important that the system is sealed during the experiment.

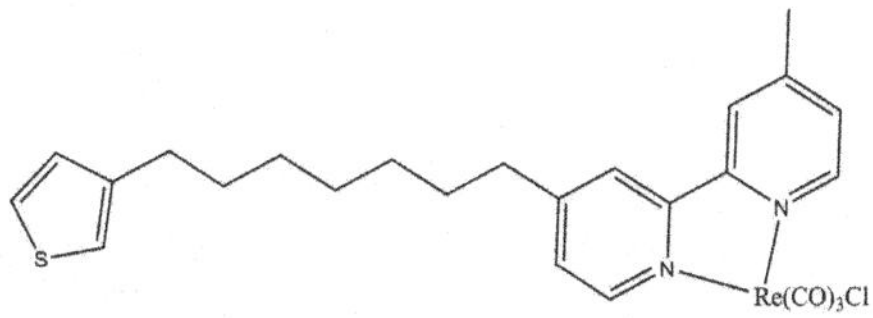

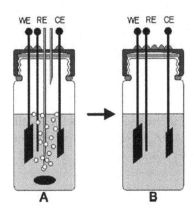

Figure 12: chemical structure of [4-Methyl-4'-(3-thienylheptyl)-2,2'-bipyridinyl)Re(CO)₃Cl]: **compound 4**

Figure 13: One-compartment cell for purging with N_2 (A) and during electropolymerization a closed system (B). Cells contain a working electrode (WE), a reference electrode (RE) and a counter electrode (CE) [39].

2.5 Calculation of the halfway-potential

For all the experiments a Ag/AgCl-quasi reference electrode is used. In order to determine the correction value for NHE a scan of Ferrocene needs to be performed. The setup, which is used, is equal to the setup used for the standard potential of the quasi reference electrode a CV using the Fc/Fc⁺ couple is recorded. 0.1 M TBAPF₆ in acetonitrile is used as electrolyte and some crystalls of Ferrocene are added. The scan range is from 0 mV to 900 mV and back to 0 mV. The halfway-potential is defined as

$$E_{1/2} = \frac{E_{red} + E_{ox}}{2}.$$

(2)

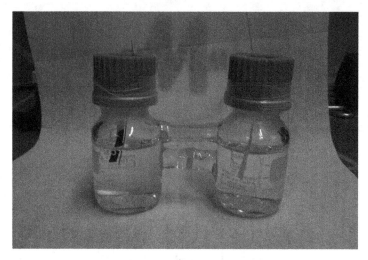

Figure 14: H-cell for electrolysis. The cell contains a WE (Pt-plate with polymeric film) and a Ag/AgCl-quasi RE on the one side of the cell and on the other side a CE (Pt-plate).

The voltammogram of Ferrocene is shown in Figure 15. According to Equation 2 the halfway potential of Ferrocene is 362 mV. The halfway potential of Ferrocene vs NHE is reported in the literature with 640 mV [41]. This means that the plots, where this Ag/AgCl-quasi RE is used, need to be shifted according to equation 3 in order to have the potential vs. NHE.

$$\Delta E = 640mV - 362mV = 278mV$$

$$(3)$$

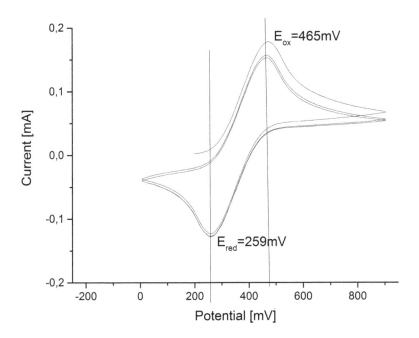

Figure 15: Cyclic voltammogram measured of Ferrocene with a scanrate of 10 mVs^{-1} in 0.1 M TBAPF$_6$ in acetonitrile.

3 Results and Discussion

3.1 Chemical Synthesis

3.1.1 Synthesis of 3-(6-bromohexyl)- thiophene

The reaction-scheme is shown in Figure 16. The product was prepared according to the reported procedure [37]. 3-Bromothiophene (16.048 g, 98.41 mmol) is dissolved in 120 mL of freshly distilled n-hexane in a three-neck roundflask and cooled down to -55°C. 40 mL (100 mmol) of 2.5 M n-buthyllithium solution in hexane are added dropwise in about 15 minutes. Afterwards 15 mL of tetrahydrofuran (THF) are added dropwise via a syringe untill the white 3-lithiothiophene salt precipitates. After 1 hour of stirring at low temperature, the mixture is let to warm up to -20°C and 60 mL (393.5 mmol) of 1,6-dibromohexane are added in one portion. The reaction is, then, let to warm up to room temperature and stirred for two hours. The solution turns yellowish during this time. The work-up consist of quenching with water followed by extraction with diethylether (3X150 mL). The organic phases are dried over Na_2SO_4 and the solvent removed by rotary evaporator. The excess of 1,6-dibromohexane was distilled out in vacuum and the residue purified by column chromatography on SiO_2 and hexane as an eluent. 5.8 g (23.4 mmol, 23.8% of yield) of pure product were isolated. ^{1}H-NMR (CDCl$_3$, 300 Mhz): δ [ppm]= 7.24 (m, 1 H of thiophene), 6.93 (m, 2 H of thiophene), 3.40 (t, J= 2.94 Hz, 2 H of hexylchain), 2.63 (t, J=7.5 Hz, 2 H of hexylchain), 1.86 (m, 2 H of hexylchain), 1.64 (m, 2 H of hexylchain), 1.36 (m, 4 H of hexylchain) (see appendix).

3.1.2 Synthesis of 4-(3-thienylheptyl)pyridine

The reaction-scheme is shown in Figure 17 and the synthesis was performed by slight modification of the literature method [38]. 4.2 mL (8.4 mmol) 2 M lithium diisopropylamine in THF/n-heptane/ethylbenzene (LDA) are diluted with 7 mL dry- THF to a three-necked roundflask cooled down to -10/-5 °C. 4-picoline (0.4 mL, 4.05 mmol) is dropwise added

23

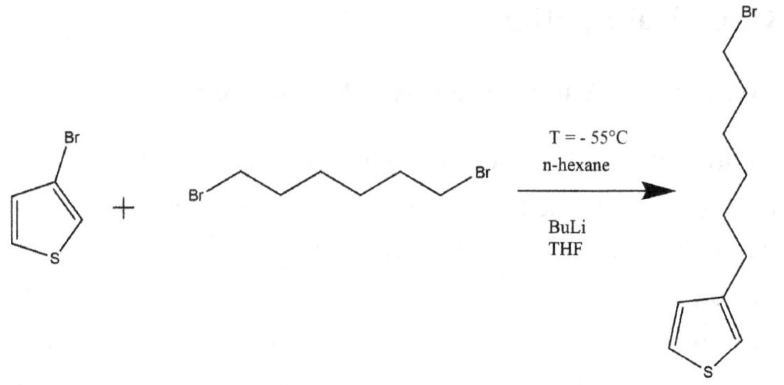

Figure 16: reaction scheme for synthesis of 3-(6-bromohexyl)thiophene

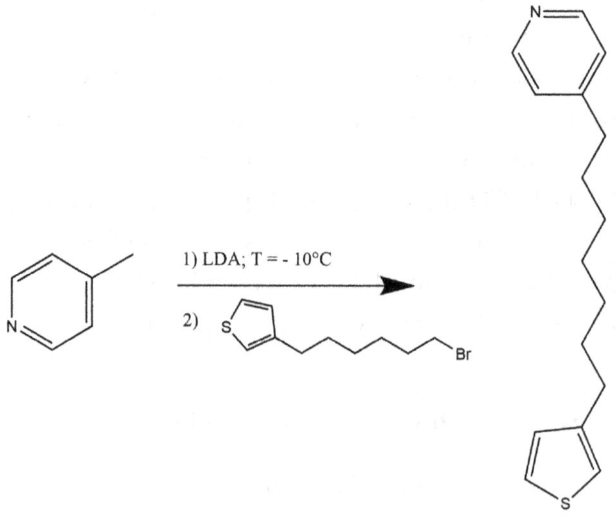

Figure 17: reaction scheme for synthesis of 4-(3-thienylheptyl)pyridine

over five minutes. The orange solution is stirred for 50 minutes. 3-(6-Bromohexyl)-thiophene (1.00 g, 4.05 mmol) diluted in 2 ml of dry THF are added dropwise over ten minutes. The solution turns from red to brown. After adding another 1 mL of THF the mixture turns pink. After stirring for 3 hours the solution turns yellow. After 5 hours the ice bath is removed, the mixture warmed up to room temperature and stirred overnight. The work-up consists of quenching the reaction with a 5% NH_4Cl-water solution at low temperature ($0°C$) and extraction with dieethylether (3x100 ml). The combined organic phases are first dried over Na_2SO_4 and the solvent is removed with rotary evaporator. 0.8 g (3.08 mmol, 76 % yield) of pure product is isolated after column chromatography on neutral alumina and a 1:1 mixutre of n-hexane: $CHCl_3$ as eluent. ^{1}H-NMR ($CDCl_3$, 300 Mhz): δ [ppm]= 8.48 (d, J=4.83 Hz, 2 H of pyridine), 7.24 (m, 1 H of thiophene), 7.09 (m, 2 H of pyridine), 6.92 (m, 2 H of thiophene), 2.60 (m, 2 H of heptylchain), 1.63 (m, 4 H of heptylchain), 1.30 (m, 6 H of heptylchain) (see appendix).

3.1.3 Synthesis of 4-Methyl-4'-(3-thienylheptyl)-2,2'-bipyridine

The reaction-scheme is shown in Figure 18 and the synthesis adapted from the reported method [38]. 0.893 g (4.85 mmol) 4,4'-dimethyl-2,2'-bipyridine are added to a three-necked roundflask, dissolved in 10 mL THF and cooled to -80°C. 2.9 mL (5.8 mmol) of 2M lithium-diisopropylamine (LDA) in THF/n-heptane are diluted in 7 mL dry THF, cooled and then dropewise added over 10 minutes. The reaction mixture is stirred for one hour. 1.413 g (5.71 mmol) 3-(6-Bromohexyl)thiophene are diluted in 5 mL dry THF and dropewise added over five minutes. The solution is stirred for one hour at -80°C. The reaction is allowed to warm very slowly to -20°C, then the cooling bath is removed and the solution is allowed to warm to room temperature. The reaction is stirred over night. The reaction is quenched with a 5% water solution of NH_4Cl and extracted with diethylether (3x100 mL). The collected organic phases are dried over Na_2SO_4 and the solvent

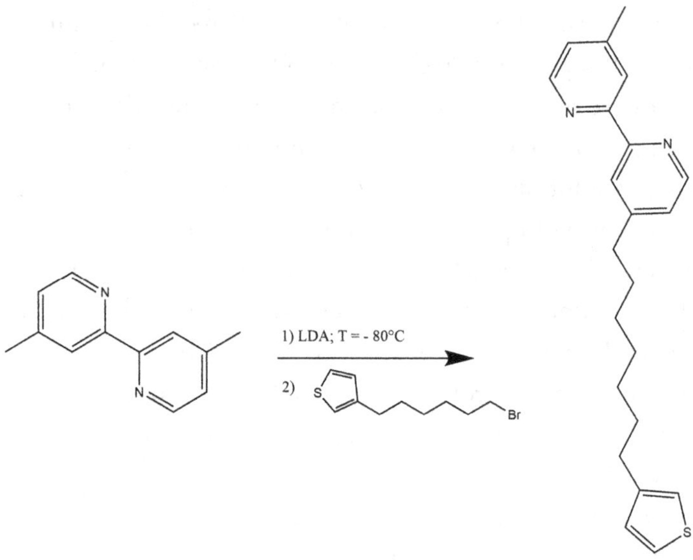

Figure 18: reaction scheme for synthesis of 4-methyl-4'-(3-thienylheptyl)-2,2'-bipyridine

removed with rotary evaporator. The crude product is 0.277g of a mixture of the free starting material 4,4'-dimethyl-2,2'-bipyridine, 4-methyl-4'-(7-(3-thienyl)heptyl)-2,2'-bipyridine and 4,4'-(bis-7-(3-thienyl(heptyl)-2,2'-bipyridine and attempt to purify the mixture through column chromatography on neutral alumina and a 1:1 mixture of n-hexane:CHCl₃ as eluent failed. The ¹H-NMR shows signals that belong to the thiophene unit (7.24 ppm (m), 6.92 ppm (m)), to the bipyridine (8.54 ppm (m), 8.22 ppm (s), 7.12 ppm (m)), to the heptyl chain (2.36 ppm (m), 1.68 ppm (m), 1.29 ppm (m)) and to the methyl group (2.43 ppm (s)) (see appendix) but it is impossible to calculate the ratio among all the possible products.

3.1.4 Synthesis of [Re(4-Methyl-4'-(7-(3-thienyl)heptyl)-2,2'-bipyridinyl)(CO)₃Cl]

The reaction-scheme is shown in Figure 19 and the synthesis was performed by adaptation of the reported methodology [40]. 0.412 g (1.14 mmol) Pentacarbonylchlororhenium(I) [Re(CO)₅Cl] are dissolved in 60 mL

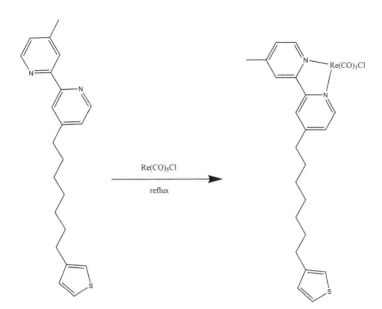

Re(CO)₅Cl

reflux

Figure 19: reaction scheme for synthesis of [4-methyl-4′-(3-thienylheptyl)-2,2′-bipyridinyl)Re(CO)₃Cl]

of dry toluene and warmed up. 0.277 g of the ligand which synthesis is descibed in section *3.1.3* are dissolved in 10 mL of dry toluene and added to the roundflask. The mixture is refluxed for one hour, cooled to room temperature and the volume of toluene reduced to about 1/10. The crude material contains the rhenium complex of 4,4′-dimethyl-2,2′-bipyridine (starting material), 4-methyl-4′-(7-(3-thienyl)heptyl)-2,2′-bipyridine and 4,4′-(bis-7-(3-thienyl)heptyl)-2,2′-bipyridine. The three products are isolated as pure substances after column chromatography on neutral alumina and a 4:1 mixture of toluene:ethyl acetate as eluent. 0.095 g (0.145 mmol) of [Re(4-Methyl-4′-(7-(3-thienyl)heptyl)-2,2′-bipyridinyl)(CO)₃Cl] were isolated. ^{1}H-NMR (CDCl₃, 300 Mhz): δ [ppm]= 8.87 (m, H_3, $H_{3'}$), 7.99 (s, $H_{1'}$), 7.95 (s, $_1$), 7.31 (m, H_2, $H_{2'}$), 7.24 (m, H_6), 6.93 (m, H_4, H_5), 2.77 (t, J=7.44 Hz, H_7), 2.63 (t, J=7.5 Hz, H_{13}), 1.70 (m, H_8, H_{12}), 1.38 (s, H_9, H_{10}, H_{11}) (see appendix).

27

3.2 Electropolymerization

Potentiodynamic electropolymerization with 0.1 M TBAPF$_6$ in acetonitrile as electrolyte, and in a second experiment using 0.1 M TBAPF$_6$ in propylen carbonate as electrolyte, were performed. As the polymerization did not work, potentiostatic electropolymerization with the same electrolytes was performed but it did not work too. The goal was to find another electrolyte or a material which lowers the oxidation potential. Borontrifluoride diethyletherate (BFEE) can be used as supporting electrolyte and it also reduces the oxidation potential of the monomer [54, 55]. The influence of this material in a scan of the [Re(4-Methyl-4'-(7-(3-thienyl)heptyl)-2,2'-bipyridinyl)(CO)$_3$Cl] monomer is shown in Figure 20. The black curve shows a scan of the monomer material in 0.1 M TBAPF$_6$ in propylencarbonate whereas the red curve shows the same scan after adding 10% of BFEE. The oxidation potential is heavily affected by adding this solvent/electrolyte. The material was only used in the polymer-growing phase of the work.

3.2.1 Electropolymerization of Pyridine functionalized poly-(thiophene) (PP1)

- Electropolymerization on Pt
  0.845 g (3.26 mmol) 4-(3-Thienylheptyl)pyridine are dissolved in 10 mL 0.1 M TBAPF$_6$ (tetrabutylammonium hexafluorophosphate) in acetonitrile. Due to the low reactivity of the monomer 10 % BFEE was added [54, 55]. Because of the high reactivity of BFEE with air, one has to flush the cell first with nitrogen for ten minutes, afterwards 1 mL (10%) of the substance is added. The potentiodynamic electropolymerization is done via cycling from 0 mV to 1800 mV back to 0 mV with a scanrate of 50 mVs^{-1}. After cycling 100 times the film is washed with acetonitrile excessively, dried and characterized.

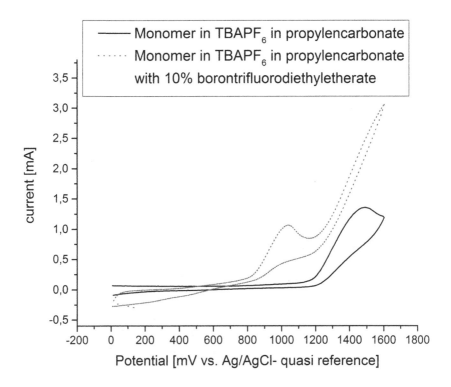

Figure 20: Scan from 0 mV to 1600 mV and back to 0 mV of [Re(4-Methyl-4'-(7-(3-thienyl)heptyl)-2,2'-bipyridinyl)(CO)$_3$Cl] monomer in 0.1 M TBAPF$_6$ in propylencarbonate and after adding 10 % of BFEE with a scanrate of 100 mVs^{-1}.

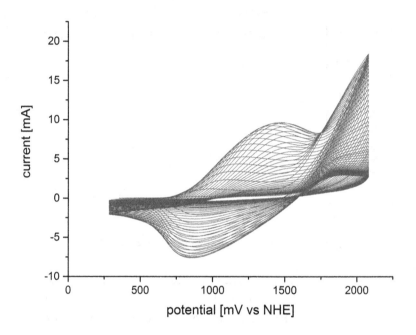

Figure 21: Potentiodynamic electropolymerization of 4-(3-Thienylheptyl)-pyridine. The polymerization is done in an one compartment-cell where a Pt-plate was used as working electrode. The counter electrode was also a Pt-plate and an Ag/AgCl-wire was used as quasi-reference electrode.

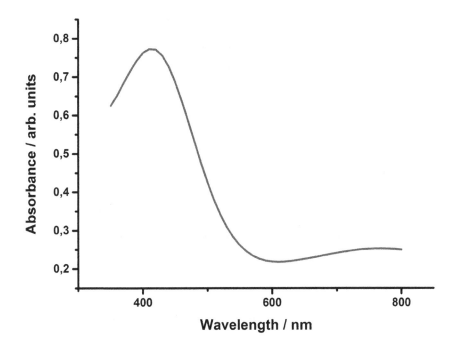

Figure 22: UV-vis measurement of pyridine functionalized poly- (thiophene) film on glass-ITO.

- Electropolymerization on a glass-ITO-plate

  The polymer is also immobilized on an glass-ITO-electrode. The polymerization is performed as described above but cycling was done for 10 cycles to get a thinner film so it can be used for an UV-vis measurement. After 10 cycles a thin green film was on the electrode. The film looked quite homogeneous and a thickness measurement with the DektakXT from bruket proved that the thickness is around 400-550 nm. Figure 22 shows the UV-vis measurement of the film.

Figure 21 shows the cyclic voltammogram during polymerization on Pt. The current increased steadily as the impedance of the electrode changed due to the growing film. The shift of the maximum of the oxidation potential indicates an α-α-linkage [35]. It is hard to start the polymerization because you have monomers but it gets easier later when you have oligomers. As there was no stirring the solution depleted of the monomer near the working electrode. Nevertheless a thick brownish film was built on the electrode.

3.2.2 Electropolymerization of 4-Methyl-4'-(7-(3-thienyl)-heptyl)-2,2'- bipyridine

0.1744 g (4.78 mmol) 4-Methyl-4'-(3-thienylheptyl)-2,2'-bipyridine are dissolved in a 0.1 M TBAPF$_6$ solution in propylen carbonate. The solution is purged with nitrogen for 15 minutes, then 1 mL of BFEE is added. The potentiodynamic electropolymerization is done with a cycling from 0 mV to 1600 mV to -500 mV with a scanrate of 50 mVs^{-1}. After cycling 30 times the film is washed with propylen carbonate and dried. After 30 cycles a thin dark green film was built on the working electrode. The cyclic voltammogram of the polymerization is shown in Figure 23.

Complexation of 2,2'-bipyridine functionalized poly(thiophene) film with pentacarbonylchlororhenium(I) (RBPP1)

0.1816 g (0.502 mmol) pentacarbonylchlororhenium(I) (ReCl(CO)$_5$) is dissolved in 25 mL dry toluene and the bipyridine functionalized poly- (thiophene) electrode is dipped in the solution for 24 h at room temperature under mild stirring as shown in Figure 24. After 24 hours stirring the solution was allowed to reflux for one hour. Afterwards the electrode was taken out, washed with propylen carbonate and dried. Neither the film nor the solution are yellow/gold coloured which would be an indication that the complexation occurred. After stirring for two hours the colour of the film started to change from dark green/brown to shiny golden- at least at some parts. The color changing stopped and just some parts of the film remained in a golden shiny color. After refluxing some parts turned white

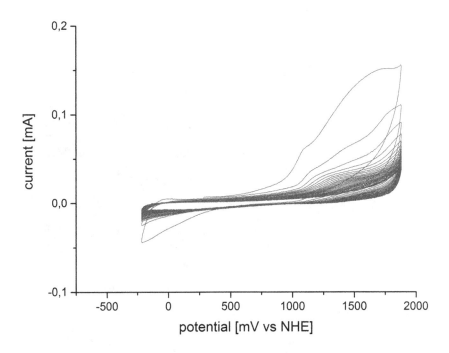

Figure 23: Potentiodynamic electropolymerization of 4-Methyl-4′-(7-(3-thienyl)heptyl)-2,2′-bipyridine with a scan rate of 50 mVs^{-1} in 0.1 M TBAPF$_6$ in propylen carbonate with 10% BFEE.

Figure 24: Complexation of poly-4-methyl-4'-(3-thienylheptyl)-2,2'-bipyridine film with Re(CO)$_5$Cl in hot toluene.

but after a while the film looked more or less like at the beginning and there was no yellow/golden color which would indicate that the complexation was successful.

3.2.3 Electropolymerization of Rhenium-bipyridine complex functionalized thiophene (RBPP2)

As the complexation after polymerization did not work the new idea is to complex first with Rhenium and electropolymerize the complex afterwards. 0.095 g (0.145 mmol) [Re(4-Methyl-4'-(7-(3-thienyl)heptyl)-2,2'-bipyridinyl)(CO)$_3$Cl] were dissolved in 15 mL 0.1 M TBAPF$_6$ in propylen carbonate. The solution is purged with nitrogen for 20 minutes. 10 % (1.5 mL) BFEE are added. The potentiodynamic electropolymerization is performed with a cycling from 0 mV to 1400 mV to 0 mV with a scan rate of 50 mVs^{-1}. After cycling 50 times the film is cleaned with propylen carbonate and dried. Figure 25 shows the cyclic voltammogram during polymerization. The current decreased as there was no stirring and the solution depleted of the monomer near the working electrode. Nevertheless, a thin, golden, shiny film was deposited on the electrode.

3.3 Characterization of the polymeric films

3.3.1 Characterization of PP1 film in acetonitrile

It is important that only the film-covered part of the Pt-plate WE is dipped into the solution, in such a way one can be sure that only the characteristics of the polymer are tested. 10 mL 0.1 M TBAPF$_6$ in acetonitrile is used as electrolyte- solution. The scan range is from 0 mV to -1800 mV and back to 0 mV and the system is always purged for 20 minutes. In order to see how fast equilibrium is reached in the system one has to make scans with different speed. First the cell is purged with nitrogen and scans with 100 mVs^{-1}, followed by a scan with 50 mVs^{-1}, as well as with 10 mVs^{-1} and a slow with 1 mVs^{-1} are done. Then the system is purged with CO$_2$ and the scans described above are repeated. Next the system is again purged

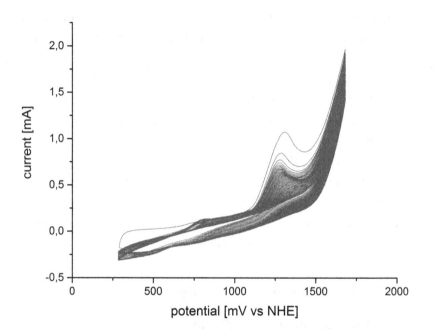

Figure 25: Potentiodynamic electropolymerization of [Re(4-Methyl-4′-(7-(3-thienyl)heptyl)-2,2′-bipyridinyl)(CO)$_3$Cl]. The polymerization is done in an one compartment-cell where a Pt-plate was used as working electrode. The counter electrode was also a Pt-plate and an Ag/AgCl-wire was used as quasi-reference electrode.

with nitrogen and a fast scan (50mVs^{-1}) is repeated to see if the process is reversible. To protonate the pyridine 0.1 mL 0.1 M sulfuric acid (H_2SO_4) are added because it was suggested in Bocarsly *et al* that the reduction process works over the pyridinium ion. As the system is still under nitrogen, the scans are repeated. Last the system is purged again with CO_2 and scaned again. If, in presence of CO_2, a reductive current is observed, an electrolysis experiment is performed to test the CO_2-reductive activity.

Figure 26 shows a cyclic voltammogram taken with a scan rate of 50 mVs^{-1}. After purging with CO_2 one can see a reductive current, which means that carbon dioxide is reduced. After acidifying the system the current increased a bit but not significantly. Figure 27 shows a slow scan with a scan rate of 1 mVs^{-1}. Here we can see a significant increase in the reductive current after acidification of the system. If you compare those two graphs one can see that the system needs some time in order to reach an equilibrium. Therefore it is always necessary to scan slowly. In the graph of the slow scan one can see that the last curve (after acidification and purging with CO_2) is not very smooth. The measurement during a slow scan is very sensitive. Still, the reductive current can be observed.

3.3.2 Characterization of PP1 film in 0.5 M KCl

15 mL of 0.5 M KCl are used as electrolyte solution. The scan rate is from 0mV to -1600 mV to 200 mV. The system is always purged for 20 minutes. In order to see how fast equilibrium is reached in the system one has to make scans with different speed. First the cell is purged with nitrogen and scans with 100 mVs^{-1}, followed by a scan with 50 mVs^{-1}, as well as with 10 mVs^{-1} and a slow with 5 mVs^{-1} are done. Then the system is purged with CO_2 and the scans described above are repeated. Next the system is again purged with nitrogen and a fast scan (50mVs^{-1}) is repeated to see if the process is reversible. Next 15 mL of 0.5 M KCl acidified with 0.1 M sulfuric acid (H_2SO_4) with an adjusted pH-value of 5.22 are purged with nitrogen and the scans are repeated. After purging again with CO_2 scans are recorded one more time. Figure 28 shows a cyclic voltammogram taken

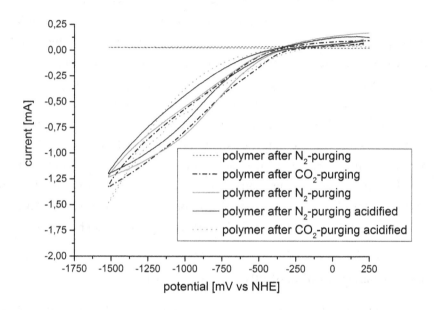

Figure 26: Cyclic voltammogram of polymeric-4-(3-Thienylheptyl)-pyridine-film measured with a scanrate of 50 mVs^{-1}. The polymerization is done in an one compartment-cell where a Pt-plate covert with the polymer was used as working electrode. The counter electrode was also a Pt-plate and an Ag/AgCl-wire was used as quasi-reference electrode. 0.1 M TBAPF$_6$ in acetonitrile is used as electrolyte.

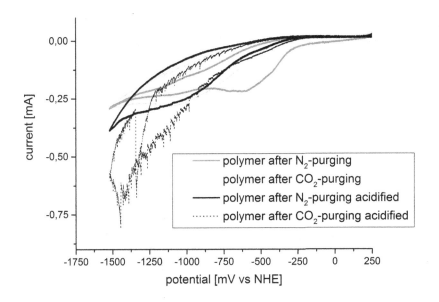

Figure 27: Cyclic voltammogram of polymeric-4-(3-Thienylheptyl)-pyridine-film measured with a scanrate of 1 mVs^{-1}. The polymerization is done in an one compartment-cell where a Pt-plate covert with the polymer was used as working electrode. The counter electrode was also a Pt-plate and an Ag/AgCl-wire was used as quasi-reference electrode. 0.1 M TBAPF$_6$ in acetonitrile is used as electrolyte.

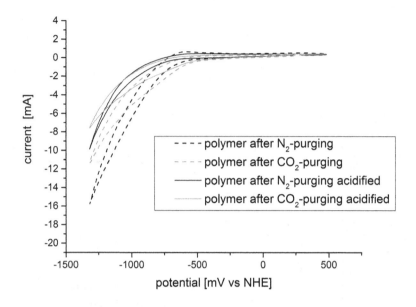

Figure 28: Cyclic voltammogram of polymeric-4-(3-Thienylheptyl)-pyridine-film measured with a scanrate of 50 mVs^{-1}. The polymerization is done in an one compartment-cell where a Pt-plate was used as working electrode. The counter electrode was also a Pt-plate and an Ag/AgCl-wire was used as quasi-reference electrode. 0.5 M KCl is used as electrolyte.

with a scanrate of 50 mVs^{-1}. Although the reductive current decreased a 40 hour electrolysis was done to be able to compare the results to those already reported in the literature [36].

3.3.3 Characterization of RBPP1 film in propylen carbonate

In order to see if the film is working for CO_2-reduction several cyclic voltammetry measurements are done. For this, a one-compartment cell is used again. 10 mL 0.1 M TBAPF$_6$ in propylen carbonate is used as electrolyte solution. The scan range is from 0 mV to -1800 mV and back to 0 mV. The system is first purged for 20 minutes with nitrogen. Scans in the described range were done with scan rates of 50mVs^{-1} and 10 mVs^{-1}.

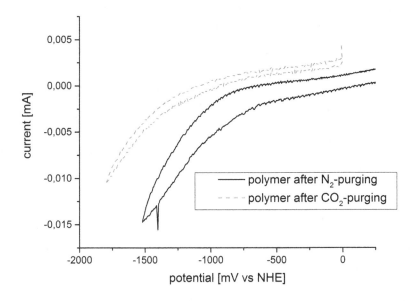

Figure 29: Cyclic voltammogram measured of the Re-complexed polymeric film with a scanrate of 10 mVs^{-1} in 0.1 M TBAPF$_6$ in propylen carbonate

The system is purged with CO$_2$ for 20 minutes and scans (same as after nitrogen-purging) are made. Figure 29 shows the cyclic voltammogram measured with a scan rate of 10 mVs^{-1} for the Rhenium-bipyridine complex functionalized poly(thiophene) film. The missing reductive current indicates that the film does not work for CO$_2$-reduction. It is quite probable that the complexation reaction did not work properly on the already made film and, for this reason, a second approach, in which the rhenium-complexed monomer is electropolymerized, is followed.

3.3.4 Characterization of RBPP2 film

In order to see if the film, which was electropolymerized after complexation, is working for CO$_2$-reduction several cyclic voltammetry measurements are perfomed. 10 mL 0.1 M TBAPF$_6$ in propylen carbonate is used as electrolyte- solution. The scanrange is from 0 mV to -2000 mV and back to 0 mV. The system is first purged for 20 minutes with nitrogen. Scans in the described range were done with scan rates of $50 mVs^{-1}$ and $10 mVs^{-1}$. The system is purged with CO_2 for 20 minutes and scans (same as after nitrogen-purging) are made. Figure 31 shows the cyclic voltammogram measured with a scanrate of $5 mVs^{-1}$ for the [Re(4-Methyl-4'-(7-(3-thienyl)heptyl)-2,2'-bipyridinyl)(CO)$_3$Cl] film. After purging with CO_2 one can see a reductive current, which means that carbon dioxide is reduced. Therefore we started a 40h electrolysis to be able to analyse the products of the reaction.

3.4 Electrolysis

3.4.1 Electrolysis of PP1 film in acidified 0.1 M TBAPF$_6$ acetonitrile solution

On each side 20 mL 0.1 M TBAPF$_6$ in acetonitrile acidified with 0.2 mL 0.1 M H_2SO_4 are used as electrolyte. The system is purged with nitrogen for 20 minutes and a relatively slow scan with $10 mV^{-1}$ is done. Then the system is purged with CO_2 for 20 minutes and a scan with $10 mV^{-1}$ is done again. A 40 hours electrolysis is done at an static potential of -1200 mV. The plot is shown in the appendix. Headspace as well as the electrolyte are analyzed after the elctrolysis. Liquid injection gas chromatography of the reaction solution diluted 1:1 with water after 40 hours electrolysis in acetonitrile containing 0.1 M TBAPF$_6$, acidified with a couple of drops of 0.1 M sulphuric acid solution in water was performed. Before injecting the solution TBAPF$_6$ is filtered off. Methanol is detected after a retention time of 2.125 minutes. For the quantitative measurement of methanol, a calibration curve in the appropriate concentration range is prepared. After 40 hours we received about $6 \cdot 10^{-8}$ mol methanol on 21.6 C. Gas chromatography as well as Cap-IC measurements showed no further reduction products.

Predicted Mechanism for the Reduction of Carbon Dioxide
Bocarsly and his group proposed a possible mechanism, which is shown in

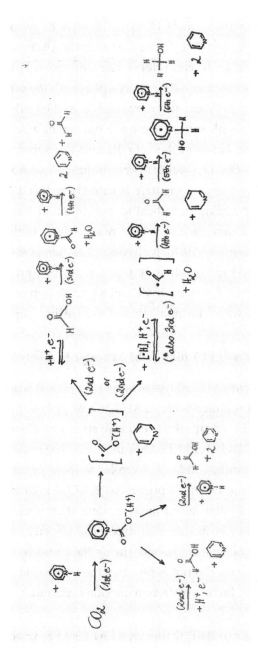

Figure 30: A proposed mechanism for CO_2-reduction with pyridinium to methanol [36].

Figure 30, for the reduction of CO_2 to methanol with pyridinium as catalyst but it is not yet proven [36]. In the first step the electroactive pyridinium cation is formed in acidic conditions. The reduction of pyridinium radical proceeds as a one electron reduction coupled with the catalytic generation of hydrogen. It is not clear if the transfer from the pyridinium radical to CO_2 occurs by a simple outer-sphere mechanism or whether it is an inner-sphere interaction between the cataylst and the substrate which would involve the formation of a radical-pyridinium-CO_2 complex intermediate. It is possible that this intermediate is directly reduced to formic acid or it reacts with a pyridinium radical to pyridine and formic acid. Formic acid may form a new radical which reacts, then, with the pyridinium radical to pyridine and formaldehyde. Another possibility is that the pyridinium-formyl radical directly reacts with the pyridinium radical to form formaldehyde. Formaldehyde and a pyridinium radical may react to another complex which forms methanol in a further step.

3.4.2 Electrolysis of PP1 film in 0.5 M water KCl solution

For this experiment an H-cell (two-compartment cell) is used. On each side 25 mL 0.5 M KCl acidified with 0.1 M H_2SO_4 at a pH-value of 5.22 are used as electrolyte. The set-up of the electrolysis is described in the previous section (Electrolysis in 0.1 M $TBAPF_6$ in ACN). The system is purged with nitrogen for 20 minutes and scans similar to those of the characterization are made. Then the system is purged with CO_2 for 20 minutes and scans are repeated. A 40 hour electrolysis is done at an static potential of -900 mV. Headspace as well as the electrolyte are analysed after the electrolysis. As the reductive current decreased during the characterisation of the film after CO_2-purging we did not expect that the film works. Product analysis showed that no reduction of carbon dioxide occurred.

3.4.3 Electrolysis of RBPP2 film in 0.1 M $TBAPF_6$ in acetonitrile

On each side of the cell 20 mL 0.1 M $TBAPF_6$ in acetonitrile is used as electrolyte. On one side of the cell the WE (Pt-plate with polymeric film,

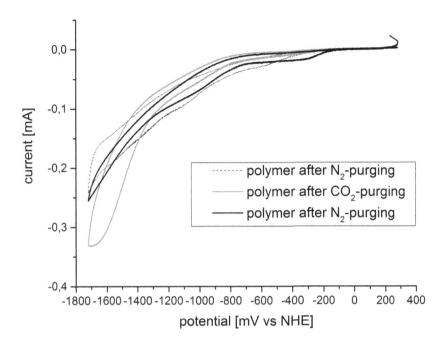

Figure 31: Characterization of [Re(4-Methyl-4'-(7-(3-thienyl)heptyl)-2,2'-bipyridinyl)(CO)₃Cl] film in 0.1 M TBAPF₆ in acetonitrile.

where only the film dips to solution) and an Ag/AgCl-wire as RE are fixed. On the other side the CE (Pt-plate), which faces the WE, is fixed. The system is purged with nitrogen for 20 minutes and a scan with 50 mVs⁻1 is done. Then the system is purged with CO_2 for 20 minutes and a scan with 50 mVs⁻1 is done again. A 40 hours electrolysis is done at an static potential of -1800 mV. The plot is shown in the appendix. Headspace as well as the electrolyte are analyzed after the elctrolysis. Carbon monoxide is detected after a retention time of 11 minutes, proving that we were able to turn an homogeneous catalyst into an heterogeneous one in this case as well. After 40 hours 250 μL carbon monoxide have been produced, which leads to a Faradaic efficiency of approximately 14%.

4 Conclusion

In this thesis the focus was on pyridine/pyridinium catalysts for methanol production and on a rhenium-bipyridine complex for CO generation. Both system have been reported to be homogeneous catalysts for CO_2 reduction in organic solvents. Within these pages those catalysts are demonstrated as heterogeneous working catalytic systems by covalently attaching them to a conductive polymer backbone (poly(thiophene)). The synthetic approaches for the pyridine functionalized and the 2,2-bipyridine functionalized compound are the same: the lithium salt of the 4-methylpyridine and 4,4-dimethyl-2,2-bipyridine reacts with 3-(6-bromohexyl)thiophene giving the pyridine, in one case, and 2,2-bipyridine, in the other, functionalized thiophene monomers. Subsequent complexation of the latter with a rhenium precursor ($[Re(CO)_5Cl]$) gives the rhenium-bipyridine functionalized thiophene monomer. Electropolymerization of the monomers in propylenene carbonate and BFFE gives functionalized films capable of reducing CO_2. This offers the opportiunity for a favorable catalytic system for CO_2-recycling to CO and CH_3OH. Further investigations will be towards different polymer-backbones and efficiency improvement . Compared to the homogeneous systems lower efficiencies were observed. However, about 0.16% Faradaic efficiency with a pyridinium functionalized poly(thiophene) and 14,29% Faradaic efficiency with a rhenium complexed bipyridine poly(thiophene) were achieved [36, 40]. Such low efficiencies are expected due to competing polymerization of the ligands with polymerization of the monomers by α-α-linkage. Gerischer *et al* proposed for the electropolymerization of various monomers an enhanced filmformation on the electrode by tuning the temperature [35]. Therefore, to improve efficiencies of the systems reported above, lowering the temperature during the electropolymerization is expected to enable more quantitative α-α-linkages and better filmformation without active site hindrance respectively. The length of the side-chain might also influence the reduction reaction as it has an influence on the quality of the polymeric film. Further it would not only be interesting to use a different monomeric unit, that

47

would be easier to electropolymerize than thiophene, but also "replacing rhenium" with other transition metals as active site for the CO_2 reduction.

Polymer	product	faradaic efficiency
Rhenium-bipyridine complex functionalized poly(thiophene)	CO	14.29%
Pyridine functionalized poly(thiophene)	CH_3OH	0.16%

5 Appendix

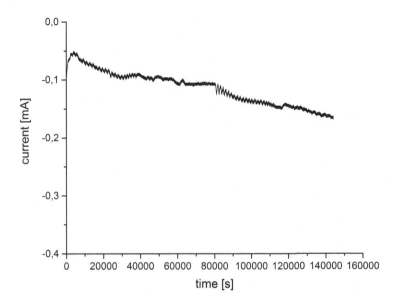

Figure 32: 40 hour electrolysis of pyridine functionalized poly(thiophene) in 0.1 M TBAPF$_6$ in acidified acetonitrile. Holdvalue at a potential of -1200 mV.

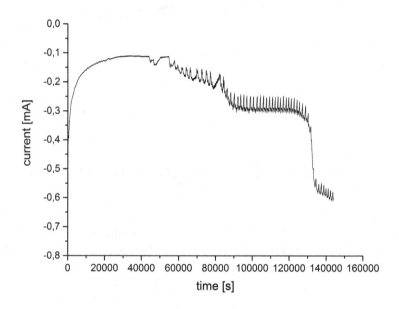

Figure 33: 40 hour electrolysis of Rhenium-bipyridine complex functionalized poly(thiophene) in 0.1 M TBAPF$_6$ in acetonitrile. Holdvalue at a potential of -1800 mV.

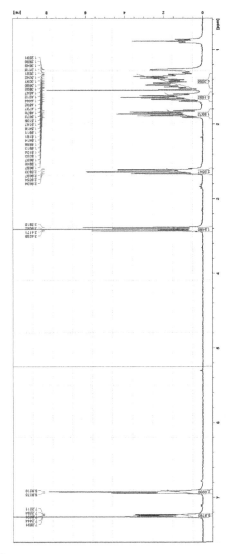

Figure 34: ^{1}H-NMR of 3-(6-Bromohexyl)-thiophene in CDCl₃

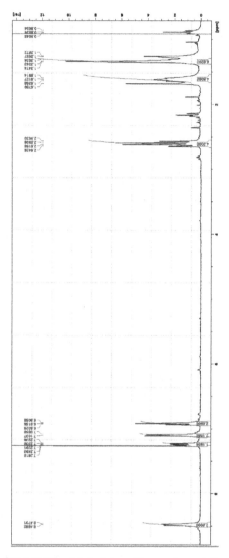

Figure 35: ^{1}H-NMR of 4-3-thienylheptyl-pyridine in CDCl$_3$

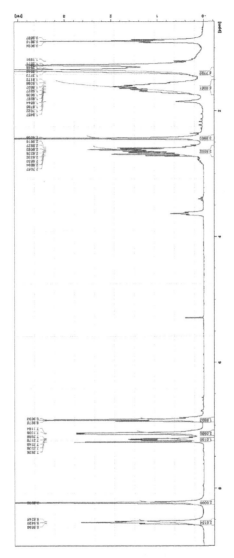

Figure 36: ^{1}H-NMR of 4'-3-thienylheptyl-2,2'-bipyridine in CDCl$_3$

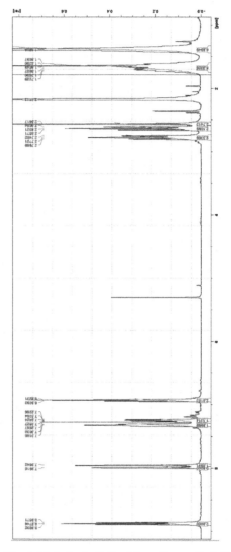

Figure 37: ^{1}H-NMR of 4-Methyl-4′-(3-thienylheptyl)-2,2′-bipyridinyl)-Re(CO)$_3$Cl

6 Bibliography

References

[1] H. Turunen. *CO_2-balance in the atmosphere and CO_2-utilisation*. An engineering approach, dissertation, Acta Universitatis Ouluensis, 2011.

[2] United States Environmental Protection Agency. *Overview of Greenhouse Gases*. http://www.epa.gov/climatechange/ghgemissions/gases/co2.html (09.04.2014).

[3] unknown. *Greenhouse effect*. http://en.wikipedia.org/wiki/Greenhouse_effect (02.05.2014).

[4] M. Lallanilla. *What Are Greenhouse Gases?*. http://www.livescience.com/37821-greenhouse-gases.html (09.04.2014).

[5] M. Aresta. *Carbon Dioxide as Chemical Feedstock*. 2010, Wiley-VCH.

[6] National Climatic Data Center, National Oceanic and Atmosperic Administration. *Greenhouse Gases*. https://www.ncdc.noaa.gov/monitoring-references/faq/greenhouse-gases.php (09.04.2014).

[7] J. M. Barnolal, D. Raynaud, Y. S. Korotkevich and C. Lorius. *Vostok ice core provides 160,000-year record of atmospheric CO_2*. Nature 329, 1987, p. 408–414.

[8] United States Environmental Protection Agency *Causes of Climate Change*. http://www.epa.gov/climatechange/science/causes.html (09.04.2014).

[9] Global Greenhouse Gas Reference Network *Trends in Atmosphheric Carbon Dioxide*. http://www.esrl.noaa.gov/gmd/ccgg/trends/index.html (09.04.2014).

[10] United States Environmental Protection Agency *Causes of Climate Change.* http://www.epa.gov/climatechange/images/science/models-observed-human-natural-large.jpg

[11] Ch. D. Keeling, R. B. Bacastow, A. E. Bainbridge, C. A. Ekdahl JR, P. R. Guenther and L. S. Waterman. *Atmospheric carbon dioxide variations at Mauna Loa Observatory, Hawaii.* Tellus XXVIII, 1976, p.538-551.

[12] S. Siitonen, L. Pirhonen. *Effects of carbon capture on gas fired power plant.* Kuala Lumpur World Gas Conference, 2012.

[13] R. S Haszeldine. *Carbon Capture and Storage: How Green Can Black Be?.* Science 325, 1647, 2009, p.1647ff.

[14] ICO$_2$N group of companies. *Carbon Capture and Storage: A Canadian Environmental Superpower Opportunity.* 2007.

[15] CESAR, Enhanced Separation and Recovery. *Cesar; Project Objectives* http://www.co2cesar.eu/site/en/about_objectives.php (15.05.2014)

[16] R. Grünwald. *Greenhouse Gas - Bury it into Oblivion. Options and risks of CO$_2$.* Technology Asessment Studies Series, No 2, 2009.

[17] J. Gibbins and H. Chalmers. *Carbon Capture and Storage.* Energy Policy 36, Elsevier, 2008, p. 4317-4322.

[18] T. F. Wall. *Combuston processes for carbon capture.* Science Direct, Proceedings of the Combstion Institute 31, 2007, p 31-47.

[19] Orion Innovations (UK) Ltd. *A UK Vision for Carbon Capture and Storage.* The Trades Union Congress and Carbon Capture and Storage Association, 2013.

[20] Ch. J. Rhodes. *Carbon capture and storage.* Science Progress 95 (4), 2012, p.473-483.

[21] H. Herzog. *Carbon Dioxide Capture and Storage.* Helm Hepburn, 2009, p. 263ff.

[22] B. Metz, O. Davidson, H. de Coninck, M. Loos and J. Meyer *Carbon Dioxide Capture and Storage*. Intergovernmental Panel on Climate Change, 2005.

[23] K. Caldeira, M. Akai et al. *Ocean storage*. IPCC Special Report on Carbon dioxide Capture and Storage, p.278-311.

[24] D. T. Whipple and P. J. A. Kenis. *Prospects of CO₂ Utilization via Direct Heterogeneous Electrochemical Reduction*. The Journal of Physical Chemistry Letters, 2010, 1, p.3451-3458.

[25] H. Herzog and D. Golomb. *Carbon Capture and Storage from Fossil Fuel Use*. Encyclopedia of Energy, Volume 1, 2004.

[26] J. Langanke, A. Wolf, J. Hofmann, K. Böhm, M. A. Subhani, T. E. Müller, W. Leitner and C. Gürtler. *Carbon dioxide (CO₂) as sustainable feedstock for polyurethane production*. Green Chem., 16, 2014, p.1865ff.

[27] S. N. Riduan. *Carbon Dioxide Fixation and Utilization*. Doctor thesis,2005.

[28] G. A. Olah. *Beyond Oil and Gas: The Methanol Economy*. Wiley-VCH Verlag GmbH und Co, 2005.

[29] Y. Lu, Z. Jiang, S. Xu and H. Wu. *Efficient conversion of CO₂ to formic acid by formate dehydrogenase immobilized in a novel alginate-silica hybrid gel*. Catalysis Today, 115 (2006), p. 263-268.

[30] S. Xu, Y. Lu, J. Li, Z. Jiang and H. Wu. *Efficient Conversion of CO₂ to Methanol Catalyzed by Three Dehydrogenases Co-encapsulated in an Alginate-Silica (ALG-SiO₂) Hybrid Gel*. Ind. Eng. Chem. Res., 2006, 45, p. 4567-4573.

[31] T. Reda, C. M. Plugge, N. J. Abram and J. Hirst. *Reversible interconversion of carbon dioxide and formate by an electroactive enzyme*. PNAS, 2008, Vol. 105, No. 31, p. 10654-10658.

[32] O. K. Varghese, M. Paulose, T. J. LaTempa and C. A. Gries. *High-Rate Solar Photocatalytic Conversion of CO_2 and Water Vapor to Hydrocarbon Fuels.* Nano Letters,2009, Vol.9, No.2, p.731-737.

[33] K. Kalyanasundaram and M. Graetzel. *Artificial photosynthesis: biomimetic approaches to solar energy conversion and storage.* Current Opinion in Biotechnology, 2010, 21, p. 298-310.

[34] G. Seshadri, Ch. Lin and A. B. Bocarsly. *A new homogeneous electrocatalyst for the reduction of carbon dioxide to methanol at low overpotential.* Journal of Electroanalytical Chemistry,372 (1994), p.145-150.

[35] H. Gerischer and Ch. W. Tobias. *Advances in Electrochemical Science and Engineering.* VCH, Vol. 1, 1990, p. 10ff.

[36] E. B. Cole, P. S. Lakkaraju, D. M. Rampulla, A. J. Morris, E. Abelev and A. B. Bocarsly. *Using a One-Electron Shuttle for the Multielectron Reduction of CO_2 to Methanol: Kinetic, Mechanistic, and Structural Insights.* JACS, 2010, 132, p.11539-11551.

[37] S. H. Liao, Y. L. Li, T. H. Jen, Y. S. Cheng and S. A. Chen *Multiple Functionalities of Polyfluorene Grafted with Metal Ion-Intercalated Crown Ether as an Electron Transport Layer for Bulk-Heterojunction Polymer Solar Cells: Optical Interference, Hole Blocking, Interfacial Dipole, and Electron Conduction.* JACS, 2012, p. 14271ff.

[38] J. Wong, M. Pappalardo and F. R. Leeme *Synthesis of 4-[ω-3-(Thienyl)alkyl]pyridines and 4-[ω-(3-Thienyl)alkyl]-2,2'-bipyridines.* Aust.J.Chem., 1995, 48, p.1425-1436.

[39] E. Portenkirchner, K. Oppelt, Ch. Ulbricht, D. A. M. Egbe, H. Neugebauer, G. Knör and N. S. Sariciftci. *Electrocatalytic and photocatalytic reduction of carbon dioxide to carbon monoxide using the alkynyl-substituted rhenium(I) complex (5,5'-bisphenylethynyl-2,2'-bipyridyl)Re(CO)₃Cl.* Journal of Organomettalic Chemistry, 716 (2012), p. 19-25.

[40] J. M. Smieja and C. P. Kubiak *Re(bipy-tBu)(CO)₃Cl-improved Catalytic Activity for Reduction of Carbon Dioxide: IR-Spectroelectrochemical and Mechanistic Studies.* Inorg. Chem., 2010, 49, p. 9283-9289.

[41] C. M. Cardona, W. Li, A. E. Kaifer, D. Stockdale and G. C. Bazan. *Electrochemical Considerations for Determining Absolute Frontier Orbital Energy Levels of Conjugated Polymers for Solar Cell Applications.* Advanced Materials, 2011, 23, p. 2367-2371.

[42] H. Herzog and D. Golomb *Carbon Capture and Storage from Fossil Fuel Use.* Encyclopedia of Energy, Volume 1, 2004, Elsevier.

[43] J. Mack and B. Endemann. *Marking carbon dioxide sequestration feasible: Toward federal regulation of CO₂ sequestration pipelines.* Energy Policy 36, Elsevier, 2010, p. 735-743.

[44] S. J. Friedann. *Carbon Capture and Storage.* UCRL-Book, 2007.

[45] C. Azar, K. Lindgrenr, E. Larson and K. Möllersten. *Carbon Capture and Storage from Fossil Fuels and Biomass- Costs and Potential Role in Stabilizing the Atmosphere.* Climatic Change 74, 2006, p.47-79.

[46] G. Centi, S. Perathoner, G. WINE and m. Gangeri. *Electrocatalytic conversion of CO₂ to long carbon-chain hydrocarbons.* Green Chem. 9, 2007, p.671-678.

[47] E. E. Benson, C. P. Kubiak, A. J. Sathrum and J. M. Smieja *Electrocatalytic and homogenous approaches to conversion of CO₂ to liquid fuels.* Chem. Soc. Rev. 38, 2009, p. 89-99.

[48] N. S. Spinner, J. A. Vega and W. E. Mustain. *Recent progress in the electrochemical conversion and utilization of CO₂.* Cat. Sci. Technol. 2, 2012, p.19-28.

[49] E. de Jong, A. Higson, P. Walsh and M. Wellisch. *Bio-based Chemicals; Value Added Products from Biorefineries.* IEA Bioenergy, Task 42 Biorefinery.

[50] T. Arai, S. Sato, T. Kajino and T. Morikawa. *Solar CO_2 reduction using H_2O by a semiconductor/metal-complex hybrid photocatalyst: enhanced efficiency and demonstration of a wireless system using $SrTiO_3$ photoanodes.* Energy Envrion.Sci.,2013, 6, p.1274ff.

[51] B. Constantz, K. Self, R. Seeker and M. Fernandez. *Carbon Capture and Storage.* United States Patent Application Publication,2011.

[52] E. E. Barton, D. M. Rampulla and A. B. Bocarsly. *Selective Solar-Driven Reduction of CO_2 to Methanol Using a Catalyzed p-GaP Based Photoelectrochemical Cell.* JACS Communications,2008.

[53] K. A. Keets, E. B. Cole, A. J. Morris, N. Sivasankar, K. Teamey, P. S. Lakkaraju and A. B. Bocarsly. *Analysis of pyridinium catalyzed electrochemical and photoelectrochemical reduction of CO_2: Chemistry and economic impact.* Indian Journal of Chemistry, Vol. 51A, 2012, p.1284-1297.

[54] J. Xu, Z. Wei, Y. Du, S. Pu, J. Hou and W. Zhou. *Low Potential Electrodeposition of High-Quality and Freestanding Poly(3-(6-Bromohexyl)Thiophene) Films.* Journal of Applied Polymer Science, Vol. 109, 2008, p. 1570-1576.

[55] H. T. Santoso, V. Singh, K. Kalaitzidou and B. A. Cola. *Enhanced Molecular Order in Polythiophene Films Electropolymerized in a Mixed Electrolyte of Anionic Surfactants and Boron Trifluoride Diethyl Etherate.* American Chemical Society, 2012, p.1697-1703.

Printed in the United States
By Bookmasters